P-26

By Larry Davis
Color by Don Greer & Tom Tullis
Illustrated by Joe Sewell

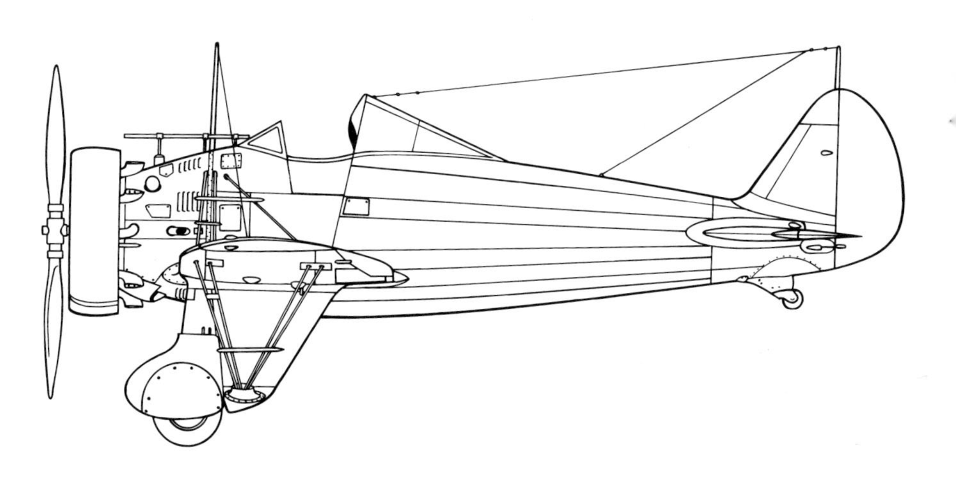

Mini Number 2

squadron/signal publications

A pair of Boeing P-26A Peashooters of the 18th Pursuit Group based at Wheeler Field, Hawaii during 1939. The aircraft in the foreground was flown by the Group Commander.

ISBN 0-89747-322-1

If you have any photographs of aircraft, armor, soldiers or ships of any nation, particularly wartime snapshots, why not share them with us and help make Squadron/Signal's books all the more interesting and complete in the future. Any photograph sent to us will be copied and the original returned. The donor will be fully credited for any photos used. Please send them to:

Squadron/Signal Publications, Inc.
1115 Crowley Drive.
Carrollton, TX 75011-5010

Acknowledgements

Air Force Museum
Vincent Berinati
Jack Binder
Peter Bowers
Bob Esposito
Jeff Ethell
David Menard
Nick Williams

The Boeing P-26 Peashooter was the last of the open cockpit, fixed landing gear, wind in the wires, pursuits to see service with the U.S. Army Air Corps. (AFM)

Introduction

The Boeing P-26, commonly known as the Peashooter, was the first mass produced monoplane fighter aircraft built by a U.S. manufacturer. It was also the last open cockpit, fixed landing gear, externally braced pursuit to serve in the U.S. Army Air Corps. The P-26 had a reasonably short operational life, only about eight and a half years, but it would become one of the best known and loved aircraft of the pre-war era.

The P-26 was the second mass-produced pursuit plane built by the Boeing Aircraft Company in Seattle, Washington. Buoyed by sales of the superb P-12/F4B biplane pursuit aircraft, Boeing's engineers set out to design and build the first monoplane pursuit aircraft for the U.S. military. Boeing already had experience in the field of monoplanes with the Army XP-9 and the model 200 Monomail, both of which were completed in the late 1920s. Although the Army was very interested in the Boeing monoplane proposals, budget constraints and military politics held up funding for the project. Two Boeing monoplane proposals, the models 224 and 245, were both rejected while still on paper. Boeing did go ahead with monoplane development, successfully selling the Y1B-9 to the Army, and various Monomail types to the airline industry.

Sales of those two types, plus continuing development and sales of the P-12/F4B kept the new military pursuit project going. In September of 1931, Boeing engineers set down on paper the basic drawings of the model 248. The Army liked what it saw, but still had no money for the project. Army and Boeing came to a mutual agreement, Boeing would design and build three model 248 prototype airframes, and Army would loan Boeing the necessary equipment to get the project off the ground. That equipment included the power plant, propeller, instruments, radio, and other military hardware necessary to complete the project. Boeing and the Army signed an agreement on 5 December 1931 to build three prototype aircraft under the Army designation XP-936.

XP-936

Construction of the first XP-936 began during January of 1932. There was so much interest in the new pursuit design that Boeing moved the engineers into the hangar bay with the construction people to avoid delays. When a part was needed, the engineers drew it up, handed the plans to the production crew, who promptly cut it or built it, and installed it on the prototype. The first XP-936 was completed in February of 1932. Going from paper to prototype in just over two weeks. Boeing Test

The first successful production venture for the Boeing Aircraft Company was the P-12/F4B series of biplane pursuits which served both the Army and Navy. (Bob Esposito)

(Above & below) The first XP-936 on the ramp at the Boeing plant in February of 1932, just after the aircraft's official rollout. The rounded wing tips and short headrest was used only on the XP-936 prototypes. The first flight of the XP-936 took place on 10 March 1932, (AFM)

Pilot Les Tower took the XP-936 into the air for the first time from Boeing Field on 10 March 1932. Tower was elated with the new pursuit's flying characteristics. It had great maneuverability and was 30 mph faster than the latest P-12. Tower delivered the XP-936 to Wright Field on 25 April 1932.

The XP-936 was twenty-three and a half feet in length and had a twenty-seven foot wingspan. The aircraft was three feet longer than a P-12, but had a three foot shorter wing span. It weighed in at 2,070 pounds empty and had a gross weight of 2,740 pounds. The XP-936 was powered by a 525 hp Pratt &

Captain Frank "Monk" Hunter, a First World War ace with six and a half victories, taxis the second XP-936 past the United Airport tower at Burbank on 10 May 1932 enroute to NAS Anacostia, then on to Wright Field where the second prototype would be used for static testing of the airframe. (AFM)

The third XP-936 over Michigan in May of 1932. The third prototype was the service test aircraft and was flown by veteran pilots of the 1st Pursuit Group at Selfridge Field under operational conditions. (AFM)

Whitney SR-1340E Wasp air-cooled radial engine. This gave the XP-936 a top speed of 227 mph, a cruising speed of 193 mph, a rate of climb of 2,260 feet/minute and a service ceiling of 27,800 feet. The fuselage was of semi-monocoque design, with brazier (raised) riveted aluminum skin. The wing was built in three sections. The center section was the load carrying section and had the landing gear attached. The outer wing panels were riveted aluminum with round wingtips. There were no underwing flaps for low speed operations.

Construction of the second and third XP-936s was completed in April and May of 1932 respectively. The second aircraft flew on 22 April, and was delivered to Wright Field, after a short demonstration flight for the Navy. It would serve as the static test airframe and was never flown again. The third aircraft went straight to the 1st Pursuit Group at Selfridge Field, Michigan, where it began flying the service test portion of the program. Even though all three prototypes were still the property of Boeing, the pilots flying the tests were all Army Air Corps officers. On 15 June 1932, the Army purchased all three XP-936 prototypes, designating them XP-26s (serial 32-412, -413 and -414). Since all three aircraft were also being used for the service test program, they were all re-designated as YP-26s during the Summer of 1932. Finally, the Army redesignated the three prototypes simply P-26s.

None of the three prototypes survived. The first aircraft served as a test and evaluation aircraft at Wright Field until it was scrapped following 465 hours of flight testing. The second prototype was static tested until it broke, being scrapped in September of 1932. The last prototype flew both test and operational flights until it crashed in October 1934 after 344 hours of flight.

The first prototype, now designated P-26, on the ramp at Wright Field during 1934 after having been retrofitted with the taller turnover structure. The aircraft has rounded wingtips and the pitot tube was on the port wing leading edge. (Jack Binder)

Development

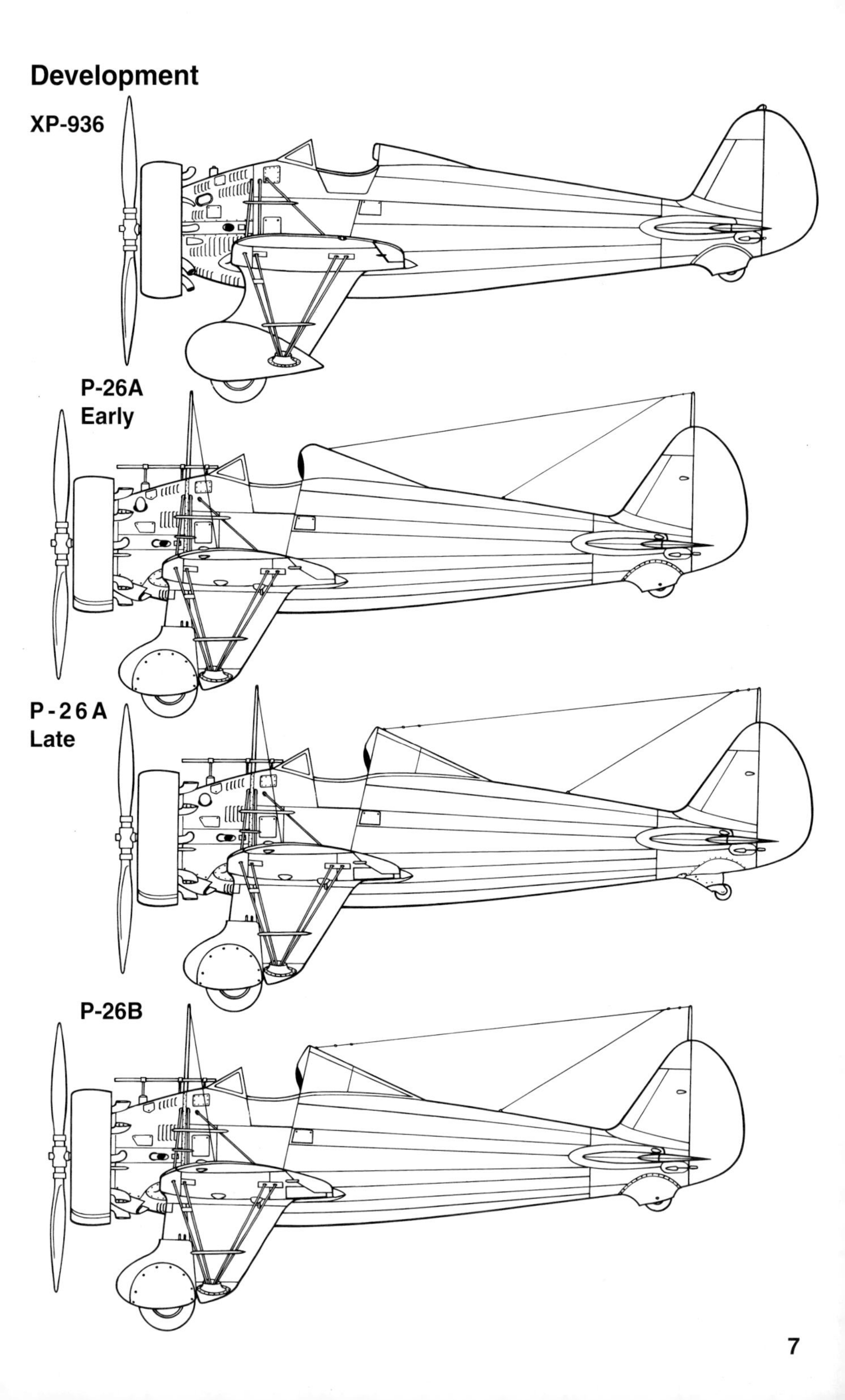

P-26A

The P-26A was the result of the Army test program with the XP-936/P-26 prototypes. Any problems encountered during the test phasc were to be resolved on the production aircraft. On 28 January 1933, the Army awarded a contract to Boeing to build one hundred eleven Boeing model 266A aircraft under the designation P-26A. Ten months after the contract was issued, on 24 November 1933, the first production P-26A (serial 33-28) was rolled out from the Seattle plant. Les Tower made the first flight on 7 December 1933, and the first operational P-26A was delivered to the 20th Pursuit Group at Barksdale Field, Louisiana the same day.

The P-26A differed from the prototypes in many ways, both internal and external. Externally the outer wing panels were redesigned with elliptical wingtips and slightly different Frise ailerons. The main landing gear wheel pants were re-designed eliminating the streamlined fairings aft of the wheels. Additionally, hand holds were added to ease the pilot entry, and the pitot tube was moved from the port wing leading edge to the starboard wing. Internally, Army-specified radios were installed, as well as floatation gear for emergency water landings.

The P-26As were the first to have armament. The armament was standard for Army Air Corps aircraft of the era - a pair of .30 caliber machine guns mounted in the lower side of the forward fuselage firing through the propeller arc. The guns were manually charged from within the cockpit. Additionally, a Type A-3 bomb rack could be fitted to the underside of the fuselage. This rack could carry up to 250 pounds of bombs.

Operational use of the P-26A brought to light many new problems, some quite hazardous. The P-26A was a very hot landing aircraft, touching down at 82 mph. The high landing speed, combined with the narrow track of the landing gear, caused many landing accidents such as ground loops. Some of these

A taller, reinforced turnover structure was fitted to all the P-26 aircraft following a turnover accident that killed a pilot from the 20th Pursuit Group. The "turnover bar" was covered by a sheet metal fairing. (AFM)

(Above/Below) The rollout of the first production P-26A (33-28) took place on 24 November 1933. The P-26A differed from the XP-936 prototype in the shape of the wingtips and wheel pants. This aircraft was the first P-26A assigned to Wright Field after the headrest modification. (AFM)

The P-26A was powered by a Pratt & Whitney R-1340-27 Wasp air-cooled radial engine and had a fixed, narrow track landing gear. Both created problems for the pilot -- the engine blocked his view and the landing gear caused many aircraft to ground loop and nose over. (AFM)

Army Air Corps pilots on the ramp at Boeing Field during 1934, ready to deliver these brand new P-26As to the 17th Pursuit Group at March Field. At the time, the P-26A was the hottest thing in the air with a top speed of 234 mph. (AFM)

ended with the P-26A flipping over onto its back. After one Barksdale pilot was killed during one of the turnovers, the Army and Boeing worked together to develop a taller and safer turnover structure. The frame of the headrest was increased in height by eight inches, and the fuselage strengthened to withstand a much higher load. The high landing speed was reduced from 82 mph to 73 mph by the development of underwing, trailing edge flaps that were hand-cranked into position. Both the headrest and flap modifications were installed on all remaining aircraft on the production line, as well as being retrofitted to all in-service aircraft.

Fuselage Development

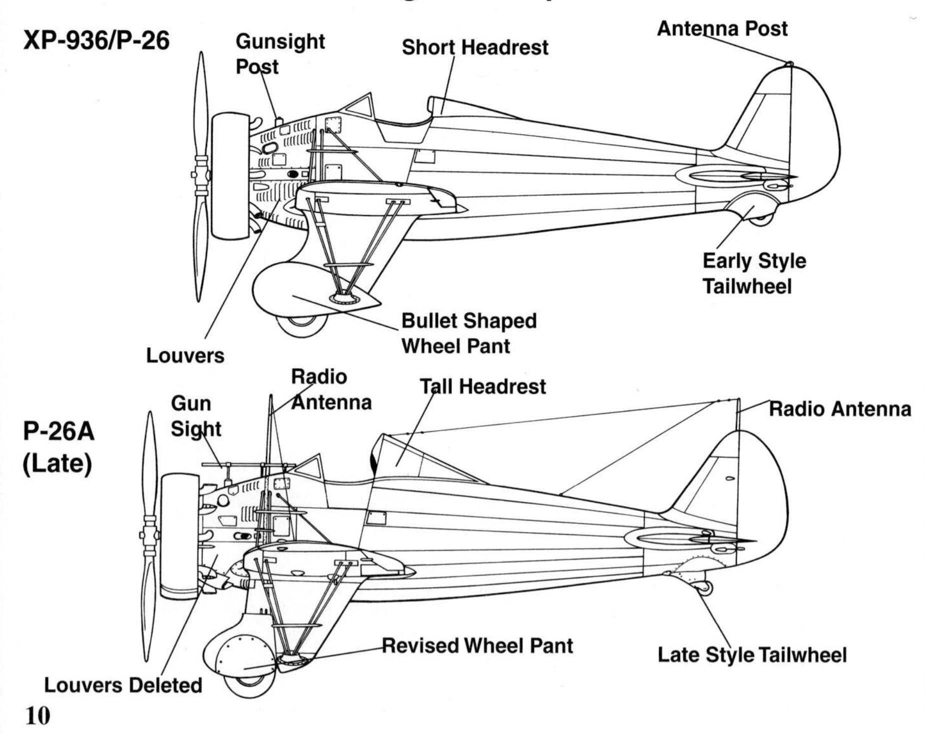

The 20th PG at Barksdale Field was the first unit completely equipped with the P-26A during early 1934. The tail code system was quite simple, P for Pursuit aircraft, T (the 20th letter in the alphabet) for the 20th Group and the individual aircraft number. (AFM)

The final design of the reinforced turnover structure fitted to the late production P-26A and retrofitted to earlier aircraft was some eight inches taller than the original turnover structure used on the XP-936. The ground crewman is opening the fuel tank to begin refueling this P-26A of the 17th Pursuit Squadron. The squadron insignia was in Black and White. (Jack Binder)

Major C.L. Tinker, commander of the 17th Pursuit Group, flies over California during 1934. The wingtips of production P-26As were elliptical in shape. The dark leading edge on the horizontal tail surfaces were rubber gravel shields, not de-icer boots. (AFM)

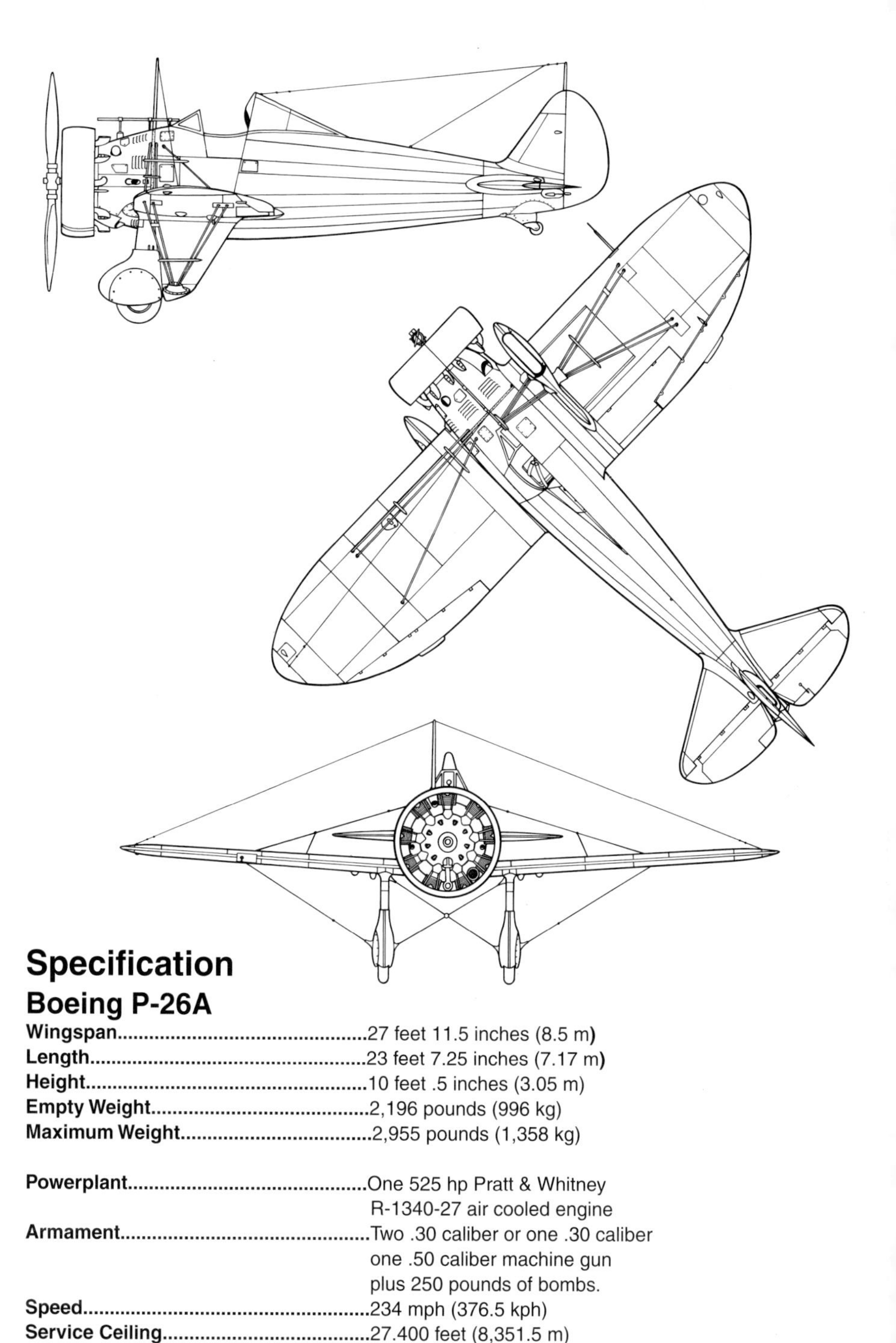

Specification

Boeing P-26A

Wingspan	27 feet 11.5 inches (8.5 m)
Length	23 feet 7.25 inches (7.17 m)
Height	10 feet .5 inches (3.05 m)
Empty Weight	2,196 pounds (996 kg)
Maximum Weight	2,955 pounds (1,358 kg)
Powerplant	One 525 hp Pratt & Whitney R-1340-27 air cooled engine
Armament	Two .30 caliber or one .30 caliber one .50 caliber machine gun plus 250 pounds of bombs.
Speed	234 mph (376.5 kph)
Service Ceiling	27.400 feet (8,351.5 m)
Range	635 miles (1,021.9 km)
Crew	One

A P-26A (33-85) of the 95th Pursuit Squadron on the ramp at March Field during 1935. Most production P-26As had provisions for radio equipment and armament, although this aircraft does not have the standard telescopic gun sight installed in front of the windscreen. (Vincent Becinati)

A P-26A assigned to the 94th PS at Selfridge Field during 1934. This aircraft has been retrofitted with an exhaust collector ring. The carburetor air intake stack is the round pipe protruding from the forward fuselage just above the open access panel. (Jack Binder)

Major Millard Harmon, commanding officer of the 20th PG at Barksdale Field, flew PT-1 during 1935 as his personal aircraft. 20th PG P-26As were unusual in that they all had the Group crest on the fuselage, rather than the more common squadron badge. The Red, Yellow and White cowl ring indicates a Headquarters Squadron aircraft. (AFM)

The problem of P-26 turnovers was so great that Boeing engineers came up with this very novel P-26 protective cockpit mockup during 1936. Although it would have protected the pilot in the event of a noseover/turnover accident, the drag was too great and this cockpit was not adopted. (AFM)

A pilot from the Bolling Field Detachment of Headquarters Command watches his instruments as the Pratt & Whitney R-1340 Wasp warms up during 1934. The Bolling Field P-26 has a Dark Blue and Yellow Townend ring cowling. The main landing gear fairings were removed to install a ski landing gear. (Peter Bowers)

The initial instrument panel for the P-26A. This panel was altered in later versions by stepping the main instrument cluster away from the back panel. The panel was finished in a Flat Black paint. (AFM)

P-26As from the 17th PS line the ramp at Selfridge Field during the Summer of 1934. All the aircraft are brand new, since the last P-26A was delivered to the 1st PG at Selfridge in June of 1934. (Fred Dickey Jr.)

This P-26A was assigned to the 94th PS before being transferred to the 17th Pursuit Group at March Field, California. It carries a Red and Yellow fuselage band, with a Red cowl ring, indicating an aircraft from the 73rd Pursuit Squadron. (J.R. Pritchard)

P-26As delivered to the 20th Pursuit Group at Barksdale Field, Louisiana, were all delivered in Gloss Olive Drab 22, with Yellow 4 wings and tails. Initial deliveries to the 20th PG had the last two digits of the serial, 33-43, painted on the fuselage side. (Vincent Berinati)

The Deputy Commander of the 18th PG at Wheeler Field, Hawaii, flew this P-26A assigned to the 19th Pursuit Squadron during March of 1939. The Townend cowl Ring and fuselage stripes are Gold, with three Black command stripes around the wings. The 18th PG was the first unit to operate the P-26A outside the continental United States. (USAF)

P-26B/C

Although the original P-26 contract called for one hundred eleven aircraft, the Army Air Corps amended that contract to include an additional twenty-five aircraft for a total of one hundred thirty-six. Two of the additional twenty-five were to be built as P-26Bs. The P-26B was identical to the late-production P-26A, except for the power plant.

The engine on the P-26B was changed from a carburetor fueled Pratt & Whitney R-1340-27 air-cooled radial to a fuel injected R-1340-33 air-cooled radial. The R-1340-33 injected engine offered an increase of 75 hp over the R-1340-27. Externally, the fuel injected versions could be easily identified by the lack of carburetor air intake stacks, which were normally seen protruding through the upper forward fuselage. However, due to the additional weight, the P-26B was only a few miles per hour faster than the P-26A. Both P-26Bs were delivered to the Army in June of 1935.

The last twenty-three aircraft on the amended contract were built as P-26Cs. They were quite simply aircraft that had all the changes called for on late production P-26As, including the flaps and higher headrest. Following a year in service, and the same amount of time for testing the fuel injected R-1430-33 engine, most of the P-26Cs (and some P-26As) were retrofitted with the R-1340 -33 engine and redesignated as P-26Bs.

Some converted P-26Cs initially had metal plates riveted over the holes in the forward fuselage where the carburetor stacks had been. The first P-26Cs were delivered to the Army in early 1936. In 1937, seventeen were converted to P-26B specifications. Several were still in service during the early days of the Second World War designated both RP-26 for Restricted flight status, and ZP-26 indicating an obsolete type.

The first P-26B was assigned to Wright Field to complete flight tests on the new Pratt & Whitney R-1340-33 fuel injected engine. Only two actual P-26Bs were built in June of 1935. The P-26B had no carburetor air intake "stacks" protruding through the forward fuselage. These stacks would have been visible just above the mechanic on the P-26A. (Jack Binder)

Fuselage Development

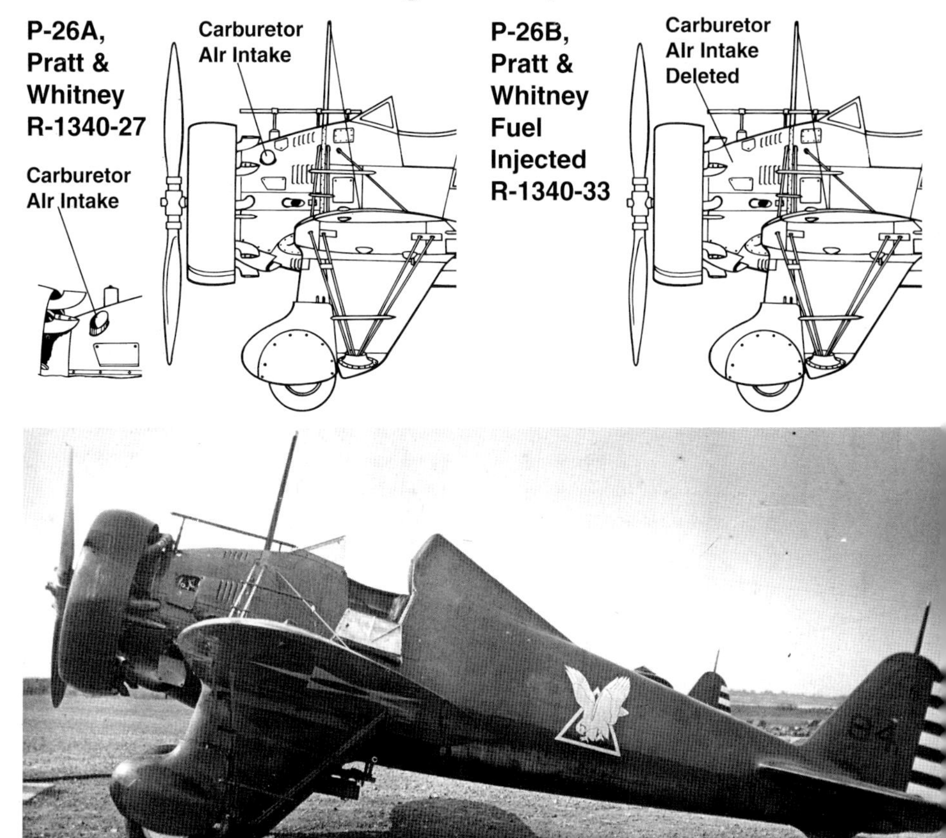

This P-26C (33-184) was assigned to the 17th PS at Selfridge Field during 1935. It has been retrofitted with the -33 fuel injected radial engine. All P-26Cs re-engined with the -33 Wasp were later re-designated as P-26Bs. (Jack Binder)

Lieutenant Colonel Ralph Royce, commander of the 1st Pursuit Group during 1936, flew this P-26C (33-181) after it had been modified with the fuel injected -33 engine and re-designated as a P-26B. It was painted standard Army Gloss Olive Drab, with Yellow wings and tail, and had a Red cowl ring. (Jack Binder)

Model 281

The Boeing model 281 was the export version of the P-26. Several countries were interested in the Boeing pursuit design, including China, which was already involved in an escalating conflict with Japan. The model 281 was identical to the initial batch of P-26Cs, with high headrest, wing flaps, and the carburetor fueled R-1340-27 Wasp radial engine. The underwing flaps were, in fact, first developed and tested on the model 281. Twelve model 281s were built, with the first one making its maiden flight on 2 August 1934. Ten were sold to China, with deliveries beginning in December of 1935. The remaining two were used as flight demonstrators. One crashed in China and the other was sold to Spain. The Chinese model 281s were heavily involved in combat with Japanese bombers over Nanking in 1936 and 1937. But by the end of 1937, attrition due to poor maintenance, lack of spare equipment, and combat, saw their removal from service. The single model 281 sold to Spain was shot down during the

The Boeing Model 281 was the export version of the P-26, and was the first variant to have underwing flaps. Only twelve Model 281s were built, with the first aircraft being rolled out painted in the standard AAC paint scheme of Gloss Olive Drab and Yellow. (Gordon S. Williams)

Smiling pilots of the Nationalist Chinese Air Force pose in front of the one of their Model 281s at Chuying Airfield during 1936. The mechanic standing on the wing is turning the hand crank used to wind the inertial starter prior to engine start. (Norman Poncetti)

(Above & Below) Ten production Model 281s were delivered to the Nationalist Chinese Air Force during early 1936. All were initially painted in overall Light Gray, with a large Black number on the side and Chinese insignia above and below both wings. (Norman Poncetti)

Wing Flaps

P-26A (Early)

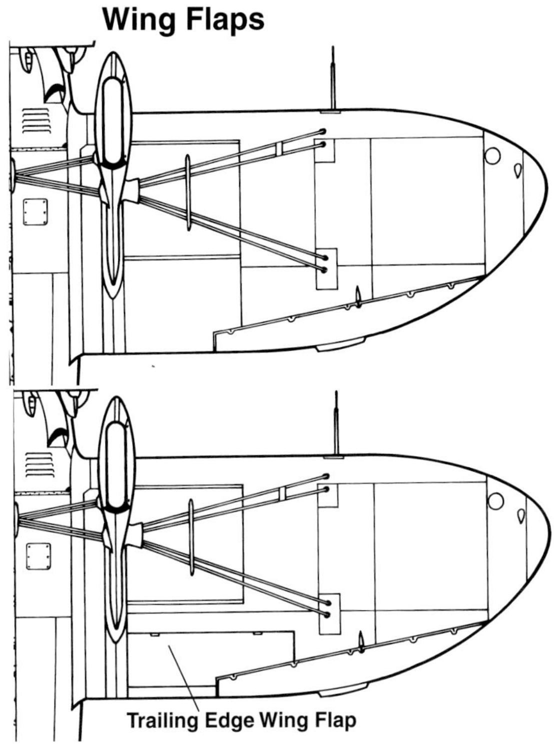

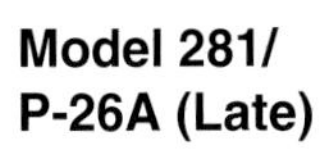

Chinese ground crewmen refuel one of the Mode1 281s at Chuying Airfield during 1936. A lack of spare parts and constant combat decimated the squadron by the end of 1937. (Norman Poncetti)

A pair of Model 281s in a hangar at Chuying Airfield during 1937. These aircraft carry the two basic paint schemes worn by Chinese Peashooters - overall Light Gray (the delivery scheme from the factory) and a camouflage scheme of overall Olive Drab. The Model 281 in the foreground was "jacked up" by attaching a rope around the propeller hub! (Norman Poncetti)

The second Model 281 (X 12775) was used as a flight demonstrator before finally being sold to Spain. It was shot down during the Spanish Civil War in October of 1936. It was rolled out at Boeing with a Gloss Black fuselage having a Red and White flash down the side, with Yellow wings and tail. (Gordon S. Williams)

In Service

The P-26 entered service with U.S. Army Air Corps units when the first P-26As were delivered to the 20th Pursuit Group at Barksdale Field, Louisiana in December of 1933. Initially three Army Air Corps groups flew the P-26A/B/C operationally - the 1st PG at Selfridge Field, Michigan; 17th PG at March Field, California; and the 20th PG. But the development of higher performance pursuits like the Seversky P-35 and Curtiss P-36, soon relegated the P-26s to second line units. In the Spring of 1937, P-26s were sent to the 3rd Pursuit Squadron at Nichols Field in The Philippine Islands. By 1940, P-26s were in service with the 37th PG protecting the Panama Canal Zone; with the 15th and 18th PGs at Wheeler Field, Hawaii; and with the 31st PG, which had taken over both the P-26 and P-35 assets previously assigned to the 1st PG at Selfridge Field.

When the Second World War broke out on 7 December 1941, there were still a number of P-26s in service in the combat areas. Many were on line at the Hawaiian bases when the Japanese struck - at least thirteen at Wheeler Field in six different squadrons. The Japanese attack destroyed six P-26s, and damaged one more. In the Philippine Islands, the initial Japanese attacks were met by a few remaining Army Air Force P-26s in the 4th Composite Group at Nichols Field. But it was the twelve P-26s in the 6th Pursuit Squadron/Philippine Army Air Corps at Batangas Field that caused the most damage.

The 6th PS/PAAC had taken over most of the P-26 assets formerly assigned to the 17th PS when that unit transitioned to P-35As in July of 1941. On 10 December 1941, the Philippine P-26 crews entered combat with the invading Japanese. On 12 December Captain

P-26As of the 20th Pursuit Group, the first group to receive the P-26A, fly a tight formation over Louisiana during 1938. The dark area under the horizontal stabilizer is a rubber stone shield. These aircraft were assigned to the 20th PG Headquarters Squadron and had the cowl rings painted in the three squadron colors; Red, Yellow, and White. (AFM)

A P-26A from the 94th PS on its back at Selfridge Field following a ground loop which resulted in a typical P-26 turnover. The reinforced turnover/headrest structure saved the pilot, although the fin and rudder were crushed (Fred Dickey Jr.)

Jesus Villamor led six PAAC P-26s against a force of fifty-four Japanese aircraft attacking near Manila. Although out-manned 9-1, the P-26s succeeded in breaking up the Japanese bomber formations, and forced the much superior force to retire. During the engagement, Villamor was credited with one bomber destroyed and several of his pilots shared in kills on two A6M Zero fighters. The PAAC force lost three P-26s. For his gallantry and leadership, Captain Villamor was awarded the U.S. Army Air Corps' Distinguished Service Cross with Oak Leaf Cluster. By the end of 1941, the Philippine P-26 force had been deci-

This very colorful P-26A was assigned to the commander of the 18th Pursuit Group at Wheeler Field, Hawaii during 1939. It carries a Gold Townend cowl ring and fuselage band, with Red, Gold and Blue stripes on the wings. (AFM)

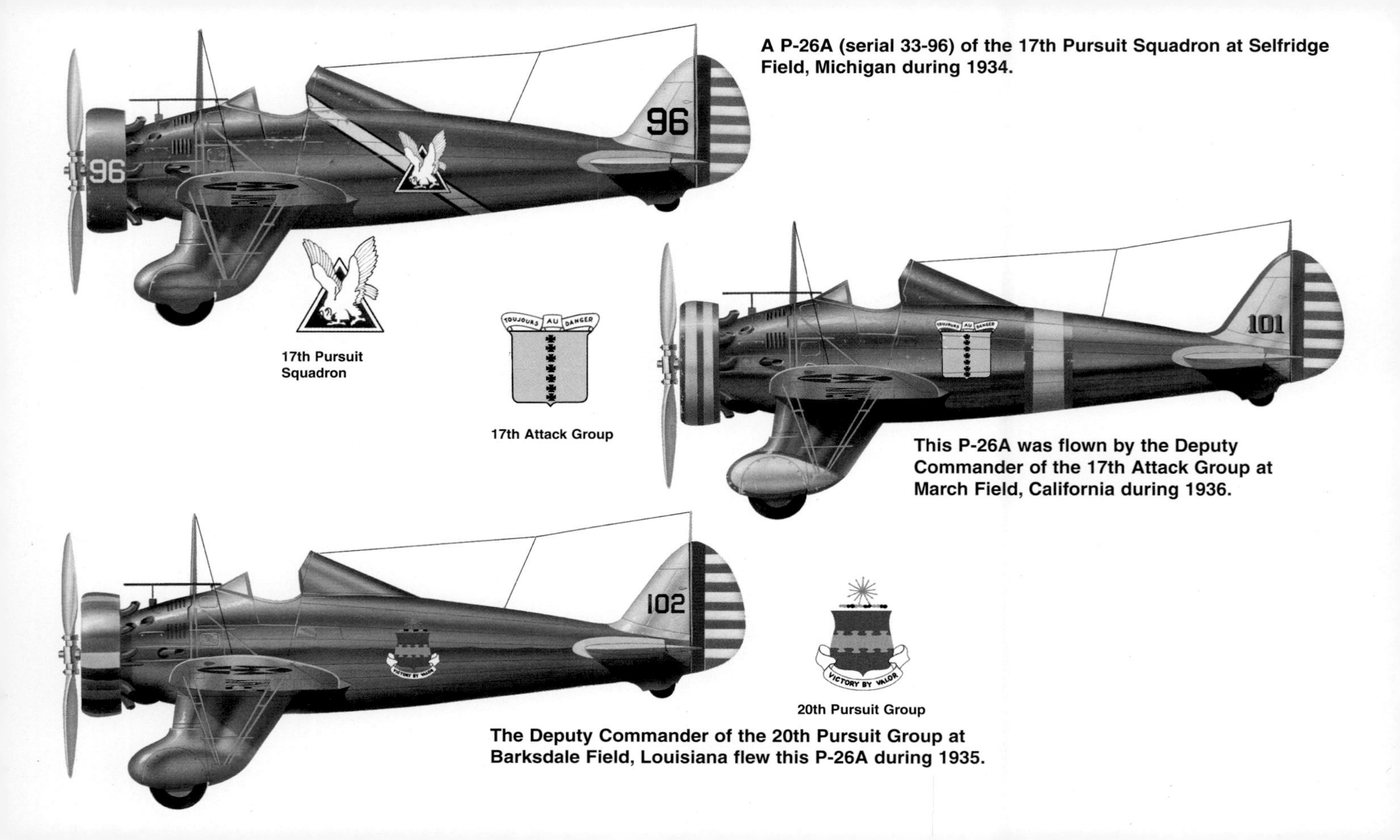

A P-26A (serial 33-96) of the 17th Pursuit Squadron at Selfridge Field, Michigan during 1934.

This P-26A was flown by the Deputy Commander of the 17th Attack Group at March Field, California during 1936.

The Deputy Commander of the 20th Pursuit Group at Barksdale Field, Louisiana flew this P-26A during 1935.

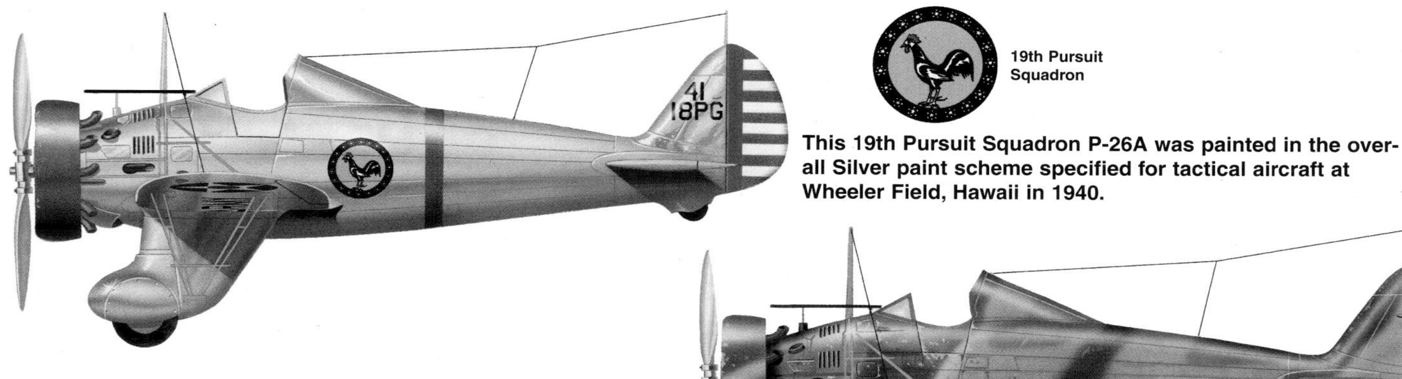

This 19th Pursuit Squadron P-26A was painted in the overall Silver paint scheme specified for tactical aircraft at Wheeler Field, Hawaii in 1940.

This P-26A of the 34th Pursuit Squadron carries a tactical camouflage of water-based paints evaluated during the 1937 War Games. During these games a number of different camouflage patterns and colors were tested

A well worn P-26A of the 67h Pursuit Squadron, Philippine Army Air Corps at Nichols Field in The Philippines on 9 December 1941. Philippine P-26s carried several different color schemes and markings combinations. One Philippine pilot, Captain Jesus Villamor, was credited with shooting down a Japanese bomber on 12 December 1941. During this engagement other PAAC pilots destroyed two A6M Zeros for a loss of three of the P-26s.

This P-26A of the 1st PG was fitted with a special tarp that covered the entire engine and cockpit during cold weather operations. Heaters could be placed under the tarp to warm the engine and cockpit before flight operations. (AFM)

mated and the remaining aircraft were burned to prevent their capture by the Japanese forces.

A number of the P-26s in service with units defending the Panama Canal Zone were sold to the governments of Panama and Guatemala. Guatemala also acquired further P-26s from Panama during the war when that nation updated its defenses with newer aircraft. Guatemala still had two P-26s in service as trainers in 1957. It was these two aircraft that were reacquired by U.S. representatives and restored and preserved for posterity. Ed Maloney's Planes Of Fame Museum acquired one of the Guatemalan aircraft, and restored it to flying condition. It was repainted in a bogus but very beautiful scheme based on the P-26s of the 34th Pursuit Squadron. The air-

All P-26 aircraft could be fitted with the Type A-9 ski installation for operations from snow covered surfaces. The skis were hydraulically raised (left) for use on dry ground and lowered (right) for use on snow. (AFM)

This 94th PS P-26A was used to test the Type A-9 ski installation. The aircraft prepares to taxi to the (dry) active runway at Selfridge Field during 1934. The Type A-9 ski was a bolt-on installation. (Warren Bodie)

craft made its "second First Flight" on 17 September 1962. The second and last remaining Guatemalan P-26 was acquired by the Smithsonian Institution, and restored by the U.S. Air Force Museum in Dayton, Ohio during 1959. The U.S. Air Force Museum displayed the completed P-26 restoration in correct 34th PS markings until 1975 when it was returned to the Smithsonian for permanent display.

P-26s of the 17st Pursuit Group share the ramp at March Field with a P-26A of the 1st PG during February of 1935. The aircraft in the foreground is the deputy group commander, the second is the group commander and the third aircraft is the commander of the 1st PG. (AFM)

The markings on 1st PG P-26s were rather mundane compared to the flashy scheme of 17th PG aircraft. This P-26A (33-65) was assigned to the 27th PS with a Yellow Townend cowl ring and tail surfaces, with Black numbers. This aircraft was equipped with the revised taller tail wheel assembly. This tail wheel assembly was a great improvement over the earlier fully faired assembly, which was easily clogged with mud on soft runways. (AFM)

Tail Wheel Development

P-26A Early

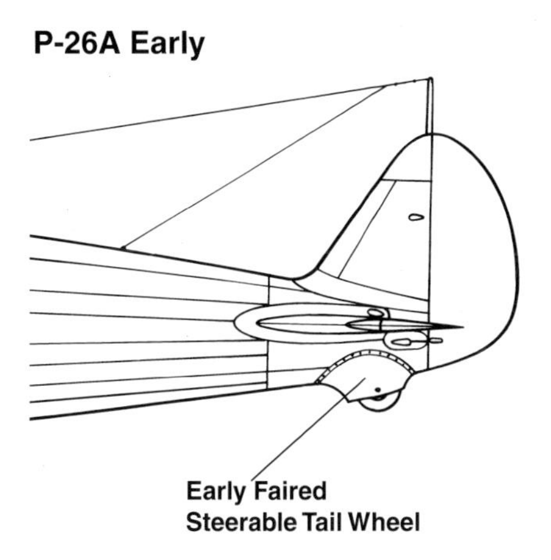

P-26A/B/C Late

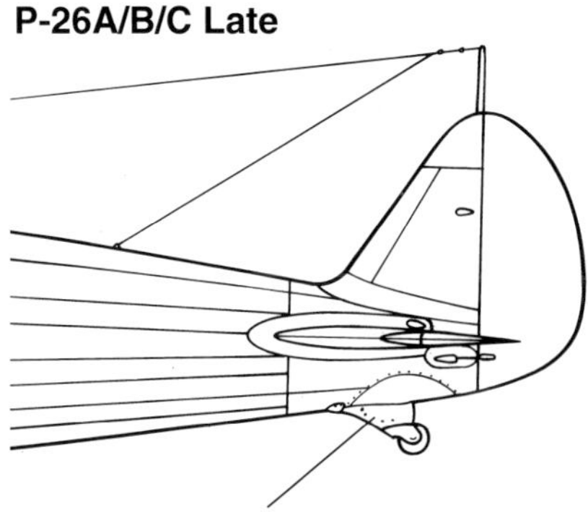

(Above & Below) In May of 1934, the Army Air Corps adopted Light Blue 23 as its primary color for all tactical aircraft, including trainers, retaining the Yellow wings and tail surfaces. These Light Blue P-26As are from the 20th Pursuit Group with aircraft 140 being from the 77th Pursuit Squadron (Red cowl ring), and aircraft 180 being attached to the 79th Pursuit Squadron (Yellow cowl ring). Their different appearance is from the type of film used which affected the Gray tone of the colors. (Jack Binder and Henry Arnold)

P-26s assigned to the 95th PS at March Field during 1935 had the scallop markings in Light Blue and Yellow on the Olive Drab fuselage, with the colors reversed on aircraft with Light Blue fuselages. The 17th PG flew P-26s for two years, 1934 and 1935, before converting back to P-12Es to train for the attack mission. Later they converted to Northrop A-17s. (AFM)

(Above/Below) Probably the most famous markings seen on P-26 aircraft are those of the 17th PG during the 1934/1935 period - in particular the 34th PS , like this P-26A (33-114). Markings on the cowl ring, fuselage, headrest, wheel pants, and tail are in Black and White. The upper surfaces of 17th PG P-26s had the standard Yellow wings, but the stabilizer was painted in the squadron colors - in this case White with Black scallops for the 34th PS. The aircraft “plane-in-squadron” number was painted on the top of the fuselage, while the squadron number (34) was painted on the underside of the fuselage. (Robert Mosher)

The 34th PS flies a tight formation over Corona, California during August of 1934. Limited Congressional funding for the Army Air Corps at this time kept aircraft contracts low and as a result, squadron strength was often less than ten aircraft per unit. (AFM)

Fourteen P-26s of the 1st Pursuit Group fly formation over Michigan during 1936. This formation included the entire 94th PS, as well as aircraft from the headquarters unit at the rear. All aircraft have the re-designed taller tail wheel structure. (AFM)

The markings of the 20th PG Deputy Commander were changed to alternating Red, Yellow and Blue stripes on the Townend cowl ring. The fuselage appears dark but is actually Light Blue 23. (AFM)

A-3 Type Bomb Rack

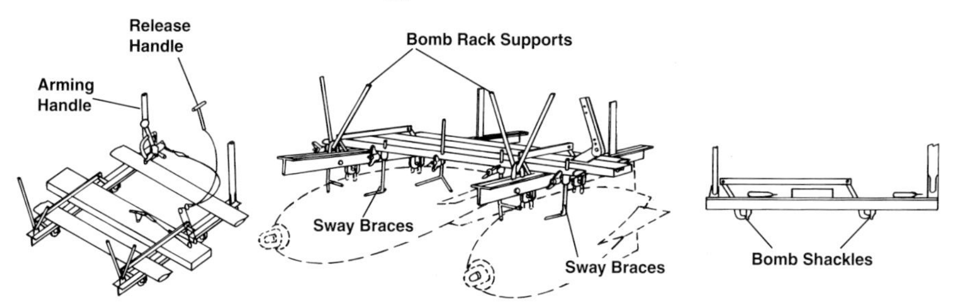

This P-26A (33-95) still carries the Light Blue and Yellow scallop markings of the 95th PS/17th PG, even though it was transferred to the 1st Pursuit Group at Selfridge Field during the Summer of 1935. The aircraft has the tail wheel fairing removed and was equipped with a bomb rack under the fuselage. (Jack Binder)

A 94th Pursuit Squadron P-26A at Selfridge Field during 1936. The aircraft's Yellow tail appears darker than the Olive Drab fuselage (due to the type of film used). The cowl ring number was Black with Yellow trim, while the fuselage band was Red with Yellow trim. The rear of the propeller blades are painted Flat Black, (Jack Binder)

The P-26As of the 73rd PS had Red and Yellow scallops on the fuselage, headrest and wheel pants, with Red tail surfaces and Yellow scallops. The serial number on top of the fuselage was also in Red with Yellow trim, as was the squadron number, 73, painted on the fuselage underside. (Peter Bowers)

A P-26C of the 17th Pursuit Squadron at Selfridge Field during 1935. The P-26C was basically a P-26A with all the various updates specified for the late production P-26A. A number of these aircraft were re-engined with fuel injected Pratt & Whitney R-1340-33 radials and redesignated as P-26Bs. (Robert Esposito)

During 1938, the 3rd Pursuit Squadron was the first unit in the 4th Composite Group at Nichols Field in The Philippines to be re-equipped with recently refurbished P-26As. The P-26A wears well maintained Gloss Light Blue 23, although the time frame was 1941. The aircraft in the background are P-35As of the 17th Pursuit Squadron. (Fred Dickey Jr.)

This P-26A (33-130) of the 1st Pursuit Group at Selfridge Field was flown by Lieutenant Harris. The aircraft was suffering from the effects of the harsh winter climate at Selfridge and has some of the Yellow paint on the vertical fin peeling off. The small crest under the windscreen carried the legend "Pilot LT Harris" and immediately under the crest is the legend "Selfridge Field." (Robert Esposito)

This P-26 is one of the P-26Cs (33-191) converted to P-26B standard with the fuel injected R-1340-33 engine. The aircraft carried no radio antenna masts when parked on the ramp at Newark Airport during 1936, even though most P-26s were radio-equipped. (Jack Binder)

P-26s of the 20th Pursuit Group fly a tight echelon formation over California. Such massed formations were common during the 1930s when the Army Air Corps was trying to keep the service and aviation in front of the public in a favorable light. Although such a formation was tactically poor, it made for good public relations (AFM)

A P-26C of the 1st Pursuit Group parked on the snow covered ramp at Selfridge Field, Michigan during February of 1936. There were a total of twenty-three P-26Cs built during 1936, most of which were later converted by Boeing to P-26B standards with fuel injected R-1340-33 engines. (Robert Esposito)

This P-26A was transferred from the Black and White scalloped 34th Pursuit Squadron to the 1st Pursuit Group. When the 17th PG became an Attack Group during March of 1935, the Army Air Corps transferred their P-26 assets to the 1st PG at Selfridge Field. Many of the aircraft retained their 17th PG markings during their early service with the 1st PG. (Jack Binder)

This pair of P-26As of the 77th Pursuit Squadron, 20th Pursuit Group at Barksdale Field, LA during 1934 carry the early style of 20th PG cowl ring markings. The cowl ring had a Red scallop with White trim and the engine facing was also painted Red. The long tube extending from in front of the windscreen was the telescopic gun sight. All P-26 variants were armed with a pair of .30 caliber machine guns mounted in the lower fuselage and firing through the propeller arc. Robert Esposito)

There is an auxiliary instrument panel just below the main instrument panel. This aircraft has had the guns removed. The "T" handle at the lower right is the gun charging handle. (AFM)

The headrest in the P-26 was made of leather. The pilot's seat was the standard Air Corps bucket, which was designed for use with a seat type parachute. The multi-pocketed item along side the seat on the cockpit wall is the pilot's map case. (AFM)

The left side of the P-26 cockpit contained the throttle, propeller, flap and engine mixture controls. The large switch on the auxiliary instrument panel is the gun control switch. (AFM)

The war games of 1934 saw many different aircraft types carrying experimental camouflage schemes applied using water-based paints. This 17th Pursuit Group P-26A has Desert Sand 26 and Neutral Gray 32 applied over the uppersurface Olive Drab fuselage and Yellow wings, with standard markings on the undersides. (AFM)

The 31st Pursuit Group was one of the pursuit units created during the Army Air Corps expansion of 1940. They were initially equipped with P-26Bs until new production Bell P-39 Airacobras became available. This 31st PG P-26B was camouflaged in Olive Drab and Dark Earth, with Neutral Gray undersurfaces. The group code and aircraft number on the fin were in Yellow. The aircraft was also equipped with a bomb rack under the fuselage. (Fred Dickey Jr.)

Colonel Ralph Royce, who had commanded the 1st PG at Selfridge Field during 1934, was the commander of the 4th Composite Group at Nichols Field in 1938 when he ground looped on landing, resulting in a typical P-26 turnover. Even though the damage appears great, the P-26 was repaired and was flyable again within a few days. (AFM)

P-26 Gun Camera Installation

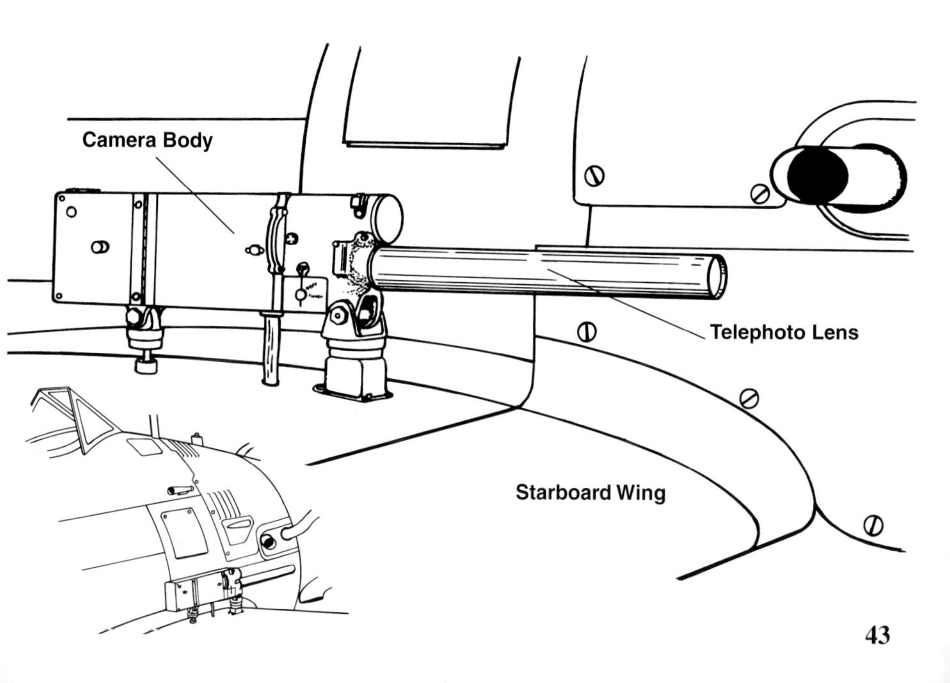

P-26As of the 1st Wing, General Headquarters Air Force, warm up their engines on the ramp at March Field, California on 15 April 1935. (AFM)

P-26s of the 94th Pursuit Squadron share a hangar at Selfridge Field Michigan, during 1937. This P-26 has been modified with the later style tall tail wheel. (AFM)

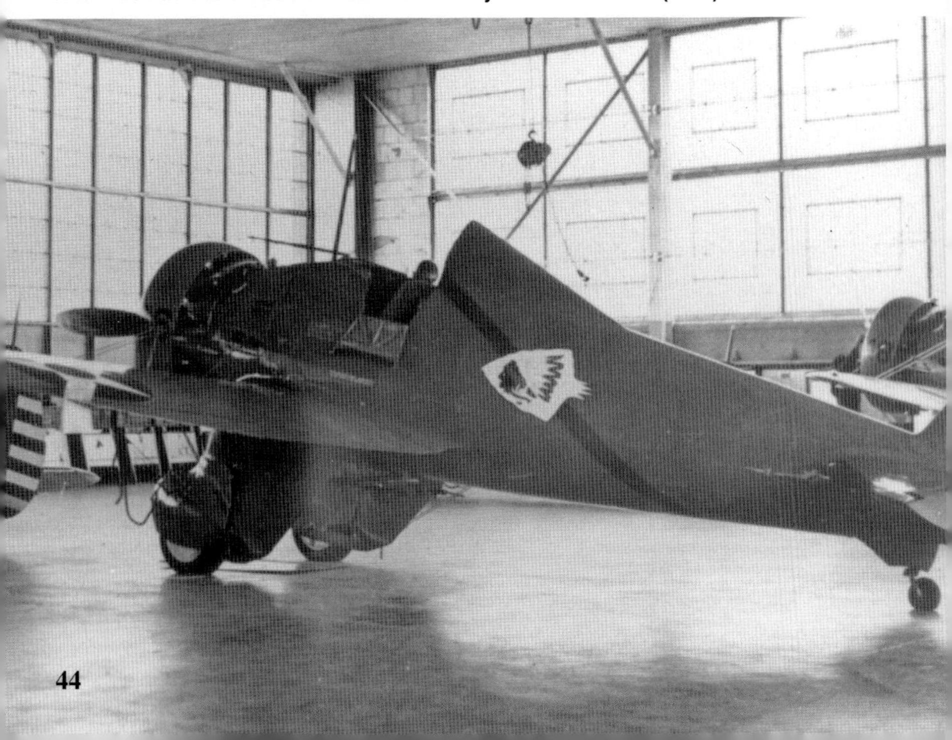

P-26 Rigging

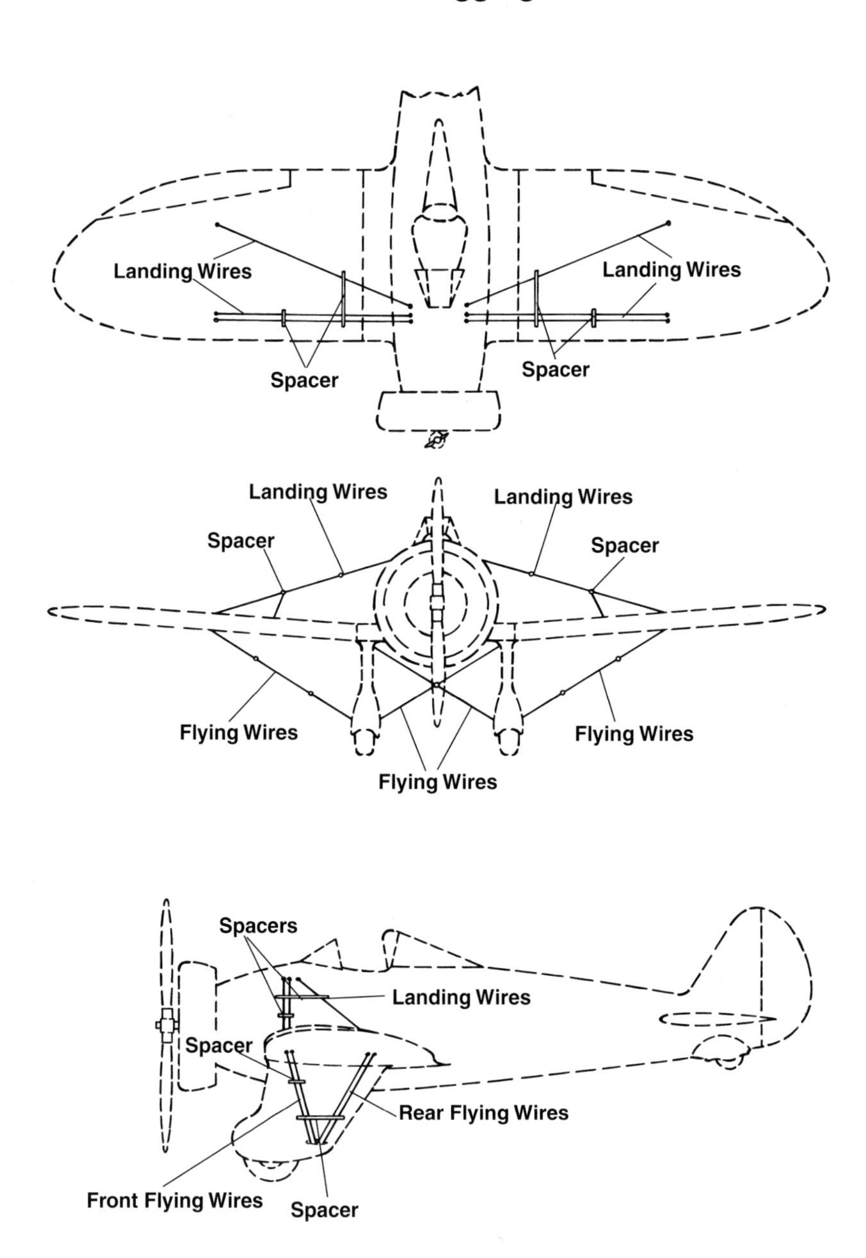

Landing Wires
Landing Wires
Spacer
Spacer
Landing Wires
Landing Wires
Spacer
Spacer
Flying Wires
Flying Wires
Flying Wires
Spacers
Landing Wires
Spacer
Rear Flying Wires
Front Flying Wires
Spacer

Beginning in 1940, Army Air Corps regulations called for tactical aircraft to be in Natural Metal or painted Silver. This Silver P-26A was assigned to the 19th Pursuit Squadron/l8th Pursuit Group based at Wheeler Field, Hawaii during late 1940. P-26s were still being used for second-line duties when the Japanese attacked in December of 1941. (T/SGT Harry Prettyman)

A P-26A from the 3rd Pursuit Squadron/4th Composite Group at Nichols Field, The Philippines. The aircraft were still painted in Light Blue 23 during late 1941. Many of the remaining P-26 assets of the 4th Composite Group were transferred to the 6th Pursuit Squadron, Philippine Army Air Corps during the Summer of 1941. (Warren Bodie)

Lieutenant Russ Church ground looped and turned over this 17th Pursuit Squadron P-26A at Nichols Field during 1941. With this type of extensive damage to the fuselage and wing, the aircraft was usually stripped for parts and scrapped. The aircraft in the background are newly arrived Seversky P-35As, (AFM)

Although all P-26s were supposed to have been turned over to the Philippine Army Air Corps, 6th Pursuit Squadron,. several Peashooters were still on the inventory of the 17th PS at Nichols Field when the Japanese attacked the Philippines. At the time, the 17th Pursuit Squadron supposedly had completely converted to the Seversky P-35A. The battle-damaged aircraft in the background are two of the unit's P-35As. It is believed that the P-26 was rebuilt from a crashed aircraft and was retained as a squadron hack. (AFM)

The radio equipment in the P-26 was rather difficult to use. The microphone was hung on an elastic cord stretched across the interior of the cockpit. This P-26A (33-65) was assigned to the 94th Pursuit Squadron at, as the legend on the fuselage states, Selfridge Field, Michigan. (James Binder)

Philippine Army Air Corps P-26s of the 6th Pursuit Squadron carried a variety of colors and markings including the Army Air Corps standard Blue fuselage with Yellow wings. Later, a number of different camouflage schemes were tried, including; uppersuface patterns of Light and Dark Earth or Dark Green and Dark Brown over Light Gray undersurfaces, overall Olive Drab and at least one aircraft was painted overall Silver. At various times, the aircraft carried U.S. and Philippine national markings, or sometimes a combination of both. This aircraft is overall Olive Drab and it still carries the USAAC insignia on the fuselage sides. (AFM)

One of the three P-26 protytypes survived the Second World War. This aircraft was used by the Aero University in Chicago during 1946 as an instructional airframe for students attending the school. (Robert Esposito)

The Guatamalian Air Force received a number of P-26As from the USAAF during the Second World War. When the aircraft were first acquired, the Guatamalian insignia was applied over the USAAF markings and the aircraft retained their Olive Drab over Neutral Gray. Later, the fuselage insignia was replaced by Black numbers on the fuselage sides. After the war, the aircraft were stripped of paint and flown in Natural Metal. Several ex-Guatamalian aircraft, including aircraft 0816, were returned to the U.S. and restored. One is owned by the Planes of Fame Museum in California and the other is at the Air Force Museum in Dayton, Ohio. (NASM)

(Above and Below) The YP-29 design evolved in the time between the appearance of the XP-936 and the first production P-26A. It shared many P-26 components and differed mainly in the wing and cockpit. The wing was a full cantilever with a rearward retracting landing gear similar to that used on the Boeing Monomail. The fuselage, tail and engine were identical to the P-26A. The Army requested that the P-29 be given an enclosed cockpit for pilot protection since the aircraft was expected to achieve speeds of over 250 mph. The first prototype had a narrow cockpit canopy which was disapproved by the Army. The aircraft was then modified with a much enlarged cockpit canopy under the designation YP-29. It was later fitted with flaps and, after service testing, was delivered to the NACA facilities at Langley Field, Virginia for use as a research aircraft. A second prototype was built with an open cockpit under the designation YP-29A and a third aircraft was produced, also with an open cockpit, under the designation YP-29B. This aircraft was sent to Chanute Field for service testing. (James Binder)